FISH FARMING (AQUACULTURE) FOR BEGINNERS

A Comprehensive Guide to Sustainable Practices, Dive into effective strategies for maximizing productivity and profit.

BRIYAN GREENWALT

Table of Contents

CHAPTER ONE 11
Introduction 11
Understanding Aquaculture: 11
Types of Fish Farming Methods: 12
Importance of Fish Farming: 13
Assessing Your Resources and Goals: 14
Choosing the Right Fish Species: 15
Understanding Legal Requirements and Regulations: 16
Planning Your Budget and Finances: 17

CHAPTER TWO 20
- Designing Your Fish Ponds or Tanks 20
  - Installing Necessary Equipment and Infrastructure: 20
  - Water Quality Management: 21
  - Stocking Your Farm with Fingerlings or Juvenile Fish: 21
- Implementing Feeding and Nutrition Strategies: 22
  - Setting up Your Fish Farm: 23
  - Installing Necessary Equipment and Infrastructure: 23

Water Quality Management: 24

Stocking Your Farm with Fingerlings or Juvenile Fish: 24

Implementing Feeding and Nutrition Strategies: 25

Fish Health and Disease Management: 25

CHAPTER THREE 26

Preventive Measures for Fish Health 26

Recognizing Symptoms of Illness in Fish: 26

Treatment Options for Diseased Fish: 27

Biosecurity Practices to Minimize Disease Risk: ...27

Food and Nutrition: 28

Understanding Fish Nutritional Requirements: 28

Selecting proper Feed kinds: 29

Feeding Frequency and Quantity: 29

CHAPTER FOUR 31

Feeding Regimes Adapted to Fish Growth Stages: 31

Supplementing Feed with Natural Food Sources:....31

Harvest and Marketing:.....32

Harvesting strategies for different fish species:.....32

Harvested fish handling and transportation: 33

Marketing methods for selling fish 34

CHAPTER FIVE 35

Environmental sustainability
.... 35

Implementing Waste Management Practices 36

Sustainable Feed Sourcing . 37

Protecting biodiversity on fish farms 38

CHAPTER SIX 39

Dealing with Predators and Pests 39

Managing invasive species: 39

Handling crises such as fish kills: 40

Addressing environmental issues from neighbors 41

Scaling up and diversifying 42

Expanding your fish farm operation: 42

CHAPTER SEVEN 43

Introducing new fish species 43

Integrating aquaponics or other complementary systems: 43

Exploring value-added goods from fish farming: 45

© 2024 [Briyan Greenwalt]. Reserved all rights.

All content in this book cannot be duplicated, shared, or conveyed in any way, including photocopying, recording, or other electronic or mechanical techniques, without the author's prior consent in writing. The only exceptions are short quotes included in reviews and certain other noncommercial uses allowed by copyright laws.

Disclaimer

The author's study, experience, and understanding of livestock management constitute the basis of the material in this book. Concerning the material contained in this book, the author is neither connected to, nor has any affiliation with, any organization, corporation, or person.

The author makes every effort to ensure that the material is accurate and thorough, but any errors or omissions, as well as any results resulting from the use of this information, are not covered by this statement. We strongly advise readers to consult a specialist for advice unique to their situation.

CHAPTER ONE
Introduction

Taking up fish farming, also known as aquaculture, opens the door to a satisfying and sustainable business activity. In this book, we'll go over the fundamentals to provide newcomers the knowledge they need to start their aquaculture journey.

Understanding Aquaculture: Simply described, aquaculture is the cultivation of aquatic organisms such as fish, crustaceans, mollusks, and aquatic plants. It is an essential strategy for addressing the increased demand for seafood while also relieving the strain on wild fish populations. Basic principles entail manipulating the environment to enhance the development, reproduction, and health of farmed species. Various systems, such as ponds, tanks, and cages, are used based on the species, region, and scale.

Types of Fish Farming Methods: Fish farming methods vary greatly, with each having its own set of pros and disadvantages.

Pond systems are the most traditional, using natural or manmade ponds to produce fish in a regulated setting. Tank systems involve raising fish in tanks or containers, which allows for exact control over water quality and the environment. Cage systems dangle fish cages in natural bodies of water, maximizing available resources yet necessitating close monitoring. Recirculating systems are highly controlled setups that continuously filter and recirculate water, increasing efficiency while reducing environmental effects.

Importance of Fish Farming: Fish farming is critical to addressing world food demand by providing a reliable source of protein and important elements. It reduces overfishing strain on wild stocks, protects natural habitats, and boosts local economies by producing jobs and promoting food security. Furthermore, fish farming enables the production of specialized species demanded by consumers, creating diversity in the seafood industry.

The benefits of starting a fish farm are considerable, both economically and environmentally. It generates a consistent income stream, particularly in locations with a high demand for fish.

Furthermore, fish farming allows farmers to diversify their revenue sources, reducing their reliance on traditional agriculture. From an environmental standpoint, fish farming can relieve pressure on overexploited wild fish populations, thereby encouraging biodiversity conservation. Furthermore, it provides chances for research and innovation in sustainable aquaculture practices, thereby improving the long-term health of aquatic ecosystems.

Assessing Your Resources and Goals: Before you begin fish farming, you must first examine your available resources and identify your goals. Consider aspects such as land availability, water supply, and financial investment. Determine your fish farming goal, whether it is for personal use, commercial selling, or environmental protection. Assessing these resources and goals will allow you to adjust your farming strategy to your individual needs and maximize success.

Choosing the Right Fish Species: Choosing the right fish species is critical for a successful aquaculture operation. Consider market demand, local climate conditions, and suitability for your farming setup.

Tilapia, catfish, and trout are popular starter selections due to their versatility and market value. Research each species thoroughly to understand their needs and growth potential, ensuring that you select the best suit for your farm.

Choosing an Appropriate Location for Your Farm: Finding the ideal location is critical to the success of your fish farm. Look for regions with access to clean water, enough land, and good environmental conditions. Consider how water temperature, pH levels, and solar exposure can affect fish health and growth. To avoid future legal complications, ensure that your chosen location conforms with zoning requirements and environmental approvals.

Understanding Legal Requirements and Regulations: Fish farming requires a variety of legal requirements and regulations that must be followed. Familiarize yourself with the local, state, and federal laws that govern aquaculture activities, such as permits, licenses, and environmental evaluations. To operate your farm legally and sustainably, make sure you follow all water consumption, waste management, and species importation rules. Consulting with regulatory agencies or legal specialists can help you negotiate these challenges more efficiently.

Planning Your Budget and Finances: Developing a comprehensive budget is critical for managing the financial parts of your fish farming business. Estimate the costs for equipment, infrastructure, fish stock, and operations expenses.

When making financial forecasts, include feed, labor, and insurance premiums. To get the finances you need for your farm, look into financing possibilities including loans, grants, or investment partnerships. Regularly assess and adapt your budget to ensure financial stability and long-term profitability.

© 2024 [Briyan Greenwalt]. Reserved all rights.

All content in this book cannot be duplicated, shared, or conveyed in any way, including photocopying, recording, or other electronic or mechanical techniques, without the author's prior consent in writing. The only exceptions are short quotes included in reviews and certain other noncommercial uses allowed by copyright laws.

Disclaimer

The author's study, experience, and understanding of livestock management constitute the basis of the material in this book. Concerning the material contained in this book, the author is neither connected to, nor has any affiliation with, any organization, corporation, or person.

The author makes every effort to ensure that the material is accurate and thorough, but any errors or omissions, as well as any results resulting from the use of this information, are not covered by this statement. We strongly advise readers to consult a specialist for advice unique to their situation.

CHAPTER TWO

Designing Your Fish Ponds or Tanks

When planning your fish farm, consider the size, shape, and depth of the ponds or tanks based on the species you intend to produce. Ensure proper drainage and access for maintenance. Select materials that are both sturdy and safe for aquatic life. Adequate distance between ponds or tanks provides for efficient management while preventing congestion.

Installing Necessary Equipment and Infrastructure: Install aeration systems to keep oxygen levels in the water, particularly in closed systems. Install filtration systems to remove waste and maintain water quality. Ensure that there is proper plumbing for water circulation and drainage. Install temperature, pH, and oxygen level sensors to track and maintain optimal fish health conditions.

Water Quality Management: Conduct regular tests on pH, ammonia, nitrites, and dissolved oxygen. Implement methods to regulate water temperature and avoid swings.

To keep your water clear, use natural or mechanical filtration methods. Keep an eye out for indicators of algae blooms or disease outbreaks, and act quickly to correct them.

Stocking Your Farm with Fingerlings or Juvenile Fish: Use reputed hatcheries or suppliers to assure high-quality, disease-free stock. To reduce stress, gently acclimate the fish to the water conditions at your farm. To prevent disease spread, quarantine new arrivals before introducing them into the existing stock. Track growth rates and change stocking densities accordingly.

Implementing Feeding and Nutrition Strategies: Choose appropriate feed formulations depending on your fish species' nutritional requirements and growth stage. Feed fish on a regular and controlled basis to avoid overfeeding and water pollution. Feeding behavior should be monitored and adjusted as needed, including feeding schedules and formulas. Consider combining commercial feeds with natural food sources such as algae or insects to stimulate natural growth while lowering expenditures.

Setting up Your Fish Farm:

Designing Your Fish Ponds or Tanks: Start by deciding on a good location for your fish farm, assuring access to water and a level surface. Ponds or tanks should be designed with the sort of fish you want to grow in mind, taking into account elements such as size, depth, and water flow. Proper planning guarantees effective space utilization and optimal fish growth.

Installing Necessary Equipment and Infrastructure: To maintain water quality and circulation in your ponds or tanks, you'll need to install equipment like aerators, filters, and pumps. Install infrastructure such as fencing and netting to safeguard your fish from predators and environmental threats. Adequate equipment and infrastructure are critical to the profitability and sustainability of any fish farming business.

Water Quality Management: To promote a fish-friendly environment, regularly evaluate water factors such as temperature, pH, oxygen levels, and ammonia concentrations. To preserve high-quality water, use measures such as water exchange, aeration, and filtration. Proper water management is critical to the health and productivity of your fish stock.

Stocking Your Farm with Fingerlings or Juvenile Fish: Purchase healthy fingerlings or juvenile fish from reliable suppliers or hatcheries. Introduce them to your ponds or tanks gradually, providing them time to adjust to their new surroundings. Monitor stocking densities to avoid overcrowding and resource competition. Careful stocking procedures lay the groundwork for a flourishing fish population.

Implementing Feeding and Nutrition Strategies: Create a feeding schedule based on your fish species' nutritional requirements and growth stages. Provide a balanced diet that includes commercial feeds, vitamins, and natural food sources. Adjust feeding frequency and quantities based on fish activity, water temperature, and development rates. Proper nutrition is critical for optimal fish development and health.

Fish Health and Disease Management: Identifying Common Fish Diseases: Familiarize yourself with common fish diseases that occur in aquaculture, such as bacterial infections, parasite infestations, and virus outbreaks. Familiarize yourself with the symptoms and diagnostic tools connected with each disease to aid in early detection and management.

CHAPTER THREE

Preventive Measures for Fish Health

To reduce the danger of disease outbreaks, implement biosecurity protocols, quarantine procedures for new stock, and conduct frequent health screenings. Maintain appropriate water quality, cleanliness, and sanitation measures to foster a stress-free environment for fish health.

Recognizing Symptoms of Illness in Fish: Keep a watchful eye on your fish for signs of illness, including aberrant behavior, hunger changes, physical abnormalities, and lesions. Conduct regular health checks to quickly detect any changes from normal fish behavior or appearance. Early diagnosis of symptoms enables timely action and treatment.

Treatment Options for Diseased Fish: As soon as you see signs of illness, separate the diseased fish to prevent disease spread throughout your farm. Consult with aquatic veterinarians or skilled aquaculturists to determine appropriate treatment choices, such as medicine, quarantine, or environmental changes. To effectively control disease, administer medicines according to the recommended dosage and duration.

Biosecurity Practices to Minimize Disease Risk: Implement biosecurity procedures to minimize the introduction and spread of infections on your fish farm. Control access to your facilities, clean equipment and vehicles, and inspect incoming inventory for illnesses. To reduce external disease sources, follow tight hygiene measures and keep the production system closed. Effective biosecurity practices are critical for maintaining the health and productivity of your fish population.

Food and Nutrition:

Understanding Fish Nutritional Requirements: Before you begin your aquaculture operation, you must first understand the precise dietary demands of the fish species you wish to raise. The dietary requirements of different fish species vary depending on their type, size, and developmental stage. By researching and understanding these requirements, you can guarantee that you are providing the proper nutrients for optimal growth and wellness.

Selecting proper Feed kinds: Once you've assessed your fish's nutritional requirements, the next step is to choose the proper feed kinds. There is a broad variety of commercial fish feeds available, each tailored to specific dietary needs. When choosing fish feed, consider protein content, fat content, and ingredient quality.

Feeding Frequency and Quantity: Setting up a feeding program is critical for maintaining healthy fish populations. Feed frequency and quantity are determined by parameters such as fish species, size, water temperature, and development rate. Feeding lesser amounts several times per day ensures optimal feed usage and reduces waste. Monitor fish behavior and adjust feeding amounts as needed to avoid overfeeding or underfeeding.

CHAPTER FOUR

Feeding Regimes Adapted to Fish Growth Stages:

As your fish grow, their nutritional needs alter. It is critical to modify feeding schedules to reflect these changes and support healthy growth. Younger fish normally require higher protein diets to facilitate rapid growth, but adult fish may require feeds with lower protein levels. Regularly monitor your fish's growth and change feeding schedules to ensure they get the nutrients they require at each stage of development.

Supplementing Feed with Natural Food Sources: While commercial feeds satisfy the majority of your fish's nutritional requirements, supplementing with natural food sources can improve their diet and resemble their natural feeding habit.

Incorporating live or natural things, such as insects, algae, or plankton, will help your fish get more nutrition and grow. Before introducing natural food sources to your aquaculture system, monitor the water quality to ensure they are safe and free of contaminants.

Harvest and Marketing:

Determining appropriate harvest times: Knowing when to harvest your fish is critical to maximizing their quality and market value. Monitoring characteristics such as growth, weight, and water temperature can assist decide the best harvest time. Regularly examining these indicators guarantees that your fish are harvested in top condition.

Harvesting strategies for different fish species: To reduce stress and ensure quality, each fish species has its own set of handling techniques during harvest. Methods such as netting, seining, and hand-picking are matched to the species' behavior and size. Understanding these strategies leads to more efficient and humane harvesting.

Harvested fish handling and transportation: Proper handling and transportation ensures that harvested fish remains in good condition from farm to market. Rapid cooling or the use of ice slurry immediately after harvest helps to prevent spoiling and ensures freshness. Fish are packaged in insulated containers with appropriate aeration to maintain quality during shipping.

Marketing methods for selling fish

Effective marketing is required to sell your fish profitably. Using internet platforms, farmers' markets, or local retailers can help you access a larger customer base. Highlighting your fish's quality, sustainability, and health benefits draws potential buyers.

Building solid ties with buyers and distributors is essential for sustaining regular sales. Providing a steady supply, high quality, and exceptional customer service develops confidence and loyalty. Regular communication and flexibility in fulfilling buyer preferences help to strengthen these connections over time.

CHAPTER FIVE

Environmental sustainability

Minimizing the environmental impact of fish aquaculture.

To reduce the environmental impact of fish farming, novices should focus on best practices such as employing recirculating aquaculture systems (RAS) to reduce water consumption and waste output. Proper site selection is critical for preventing habitat damage and reducing pollution. Furthermore, using effective feeding strategies and routinely checking water quality can help maintain a healthy environment for both fish and the surrounding ecosystems.

Implementing Waste Management Practices

Beginners in fish farming can effectively control waste by using methods such as biofiltration and sedimentation ponds to remove excess nutrients and organic debris from the water.

Organic waste can be used as fertilizer for crops or composted to lessen its environmental impact. Proper solid waste disposal and regular repair of farming equipment are required to prevent contamination and keep the environment clean.

Sustainable Feed Sourcing

Finding sustainable feed is critical for environmentally responsible fish farming. Beginners should prioritize feeds prepared with responsibly sourced fishmeal and plant-based ingredients to limit their reliance on wild fish sources. Exploring alternate protein sources, such as algae and insect meal, can also help to create a more sustainable feed chain. Furthermore, reducing feed waste and streamlining feeding techniques can boost sustainability efforts.

Sustainable fish farming requires the conservation of water resources. Water recycling and rainwater collection are two ways that beginners can use to reduce their freshwater consumption. Installing water-saving technologies and maintaining correct water circulation systems can also help to improve water consumption efficiency. Monitoring water quality measurements and preventing pollution are critical components of preserving water resources and maintaining the health of aquatic ecosystems.

Protecting biodiversity on fish farms

To encourage biodiversity conservation in fish farms, newcomers should design habitats that resemble natural environments, such as aquatic vegetation and shelter structures. Avoiding the use of toxic chemicals and pesticides contributes to a healthy ecosystem.

Introducing native species and minimizing the introduction of non-native species into nearby waters are critical steps toward preserving local biodiversity.

CHAPTER SIX
Dealing with Predators and Pests

Installing physical barriers around the fish farm, such as fences or nets, is an efficient way to repel them. Furthermore, adding natural pest predators, such as particular kinds of birds or insects, can aid in pest population control while minimizing environmental impact. It is critical to monitor the farm regularly for indicators of predators or pests to respond quickly if necessary.

Managing invasive species: Preventing the introduction of invasive species is critical in fish farming. Implementing strong biosecurity procedures, such as sanitizing equipment and monitoring water sources, can assist in reducing the danger of importing invasive species onto the farm. Regular examination and maintenance of equipment can also aid in the detection and removal of any invasive species that may have mistakenly infiltrated the farm.

Handling crises such as fish kills: In the event of a fish kill, it is critical to rapidly identify the source and take appropriate measures to prevent further losses. Low oxygen levels, illness outbreaks, and water contamination are all potential causes of fish fatalities. Regular monitoring of water quality measurements can aid in the early detection of problems. Using emergency aeration systems and backup power sources can assist keep oxygen levels stable during power outages or other situations.

Addressing environmental issues from neighbors

Maintaining open communication with nearby communities is essential for addressing any environmental concerns they may have about the fish farm. Implementing best management practices, such as efficient waste disposal and noise and odor reduction, can assist relieve problems. Offering educational outreach events or facility visits can also help to build community knowledge and support.

Ethical handling of fish entails providing proper living conditions and limiting stress throughout their life cycle. This includes keeping water clean, giving enough room, and guaranteeing correct nutrition. Regular health checks by experienced professionals can aid in the early detection and resolution of any concerns. Furthermore, using humane harvesting procedures, such as stunning before processing, guarantees that fish are handled ethically throughout the whole agricultural process.

Scaling up and diversifying

Expanding your fish farm operation: Begin by examining your current infrastructure and resources. Consider expanding your ponds or tanks to accommodate additional fish. To maintain optimal growing conditions, ensure that you have enough amount and quality of water, as well as proper filtration and aeration systems. Additionally, prepare for the additional workload and, if required, consider employing more employees.

CHAPTER SEVEN
Introducing new fish species

Before adding new fish species to your farm, research their specific water quality, temperature, and food needs. Choose species that are appropriate for your current setup and climate circumstances. To reduce stress and increase survival rates, gradually adapt young fish to their habitat. Monitor their growth and health closely to ensure they thrive in their new environment.

Integrating aquaponics or other complementary systems: Adding aquaponics to your fish farm can increase efficiency and output by using fish waste to fertilize plants, which then filter and clean the water for the fish. Begin by establishing a small-scale aquaponics system, then gradually grow as you acquire experience. Experiment with different plant species to see what grows best in your system, and consider including other complementing systems like hydroponics or vermiculture to improve resource usage.

Collaborating with other fish farms or researchers can provide useful insights, resources, and assistance to help you improve your farm's operations. Attend industry events, workshops, and conferences to network with colleagues and stay up to current on the newest developments in aquaculture. Form partnerships or join cooperative groups to address difficulties, share information, and obtain money for research and development projects.

Exploring value-added goods from fish farming: Diversifying your product offers will help your fish farm become more profitable and marketable. Consider value-added goods like smoked fish, fillets, fish jerky, and fish-based spreads and dips. Experiment with different recipes and packaging options to develop distinctive items that appeal to your target market. Invest in branding and marketing to effectively promote and differentiate your value-added items from competitors on the market.

Finally, fish farming, often known as aquaculture, provides a viable entry point for newcomers interested in agriculture and food production. Aquaculture, with its numerous species and methods, provides a versatile and scalable potential for both individuals and communities.

First and foremost, fish farming meets the growing need for seafood sustainably. As wild fish supplies decline owing to overfishing and environmental degradation, aquaculture offers a way to supply demand without further diminishing natural resources. Farmers can reduce their impact on marine ecosystems and help conserve aquatic biodiversity by raising fish in regulated surroundings.

Furthermore, fish farming provides economic benefits to newcomers looking for business alternatives. With adequate design and administration, aquaculture enterprises can produce consistent revenue streams from the sale of fish and related items. Whether using small-scale backyard ponds or bigger commercial facilities, ambitious farmers can adjust their operations to their resources and objectives.

Furthermore, fish farming promotes creativity and technological breakthroughs in agriculture. From modern water filtration systems to automated feeding mechanisms, the sector is always evolving to increase efficiency and sustainability. Beginners in the field have access to a multitude of information and resources to help them improve their techniques and increase production.

Fish farming also contributes significantly to global food security by diversifying the protein sources available to populations around the world. As populations rise and land-based resources become more scarce, aquaculture presents a feasible option for meeting dietary needs and alleviating hunger.

Finally, fish farming has enormous potential for beginners interested in learning more about agriculture and aquaculture. Aspiring fish farmers can contribute to food supply, environmental protection, and economic growth on a local and global scale by stewarding resources responsibly and implementing new approaches.

www.ingramcontent.com/pod-product-compliance
Lightning Source LLC
Chambersburg PA
CBHW072019230526
45479CB00008B/302